3-4歲

幼兒全方位
智能開發

英文篇

Let's Learn ABC
小楷字母

園丁文化

Small Letter a
小楷字母 a

aeroplane 飛機

arm 手臂

● Circle the letter "a"s. 請圈出小楷字母 a。

● Let's write the letter a. 齊來寫一寫。

a	a			

2

Small Letter b
小楷字母 b

banana 香蕉

bear 熊

● Draw a line to match the picture with the correct word.
請把圖畫和正確的詞彙用線連起來。

1.

• butterfly •

• boat •

2.

● Let's write the letter b. 齊來寫一寫。

b	b			

答案：1. butterfly；2. boat

3

Small Letter c
小楷字母 c

cow 牛

cake 蛋糕

● What do you put on a birthday cake? Put a "✔" in the correct box.
你會在生日蛋糕上插上什麼？請在正確的 ☐ 內加 ✔。

☐ candle　　☐ crayon　　☐ carrot

● Let's write the letter c. 齊來寫一寫。

c	c				

答案：candle

4

Small Letter d
小楷字母 d

duck 鴨

doctor 醫生

● Help the dog find its way home by colouring the letter "d"s.
  請把字母 d 填上顏色，帶小狗回家吧。

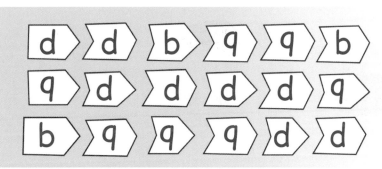

d	d	b	q	q	b
q	d	d	d	d	q
b	q	q	q	d	d

HOME

● Let's write the letter d. 齊來寫一寫。

d	d			

5

Small Letter e
小楷字母 e

 **eye** 眼睛

 **ear** 耳朵

● Who has the biggest ears? Circle the correct picture.
誰的耳朵最大？請圈出正確的圖畫。

1. 　　2. 　　3.

● Let's write the letter e. 齊來寫一寫。

e	e				

答案：3. elephant

6

Let's review a to e
溫習 a 至 e

● The bees are busy working! Match the bee with the flower with the same letter.
   蜜蜂正在忙碌地工作。請用線把寫有相同字母的蜜蜂和花連起來。

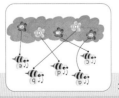

答案：

Small Letter f
小楷字母 f

frog 青蛙

fire 火

● Join the fish in the pond to spell the word "fish".
請把下面的魚依 f-i-s-h 的順序用線連起來。

● Let's write the letter f. 齊來寫一寫。

f	f			

Small Letter g
小楷字母 g

goat 山羊

grass 草

● Circle the letter "g"s.　請圈出小楷字母 g。

● Let's write the letter g.　齊來寫一寫。

g	g			

9

Small Letter h
小楷字母 h

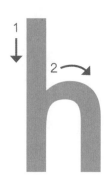

house 房子

hand 手

● What is this? Circle the correct word. 這是什麼？請圈出正確的詞彙。

hippopotamus   hen

heart   helicopter

● Let's write the letter h. 齊來寫一寫。

h	h			

答案：helicopter

10

Small Letter i
小楷字母 i

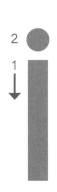

ink 墨水

ice cream 雪糕

● Join the dots with the letter "i"s. What is this? Fill in the letter "i".
請把所有 i 用線連起來，看看這是什麼，然後在橫線上填寫小楷字母 i。

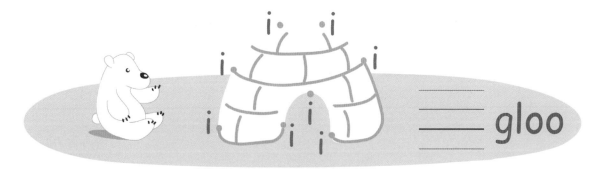

gloo

● Let's write the letter i. 齊來寫一寫。

i	i			

11

Small Letter j
小楷字母 j

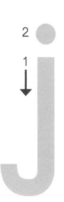

juice 果汁

jam 果醬

● Fill in the letter "j". 請填寫小楷字母 j。

＿elly

＿ar

● Let's write the letter j. 齊來寫一寫。

j	j				
j	j				

12

Let's review f to j
溫習 f 至 j

● Match each letter with the correct word and picture.
請用線把字母和正確的詞彙及圖畫連起來。

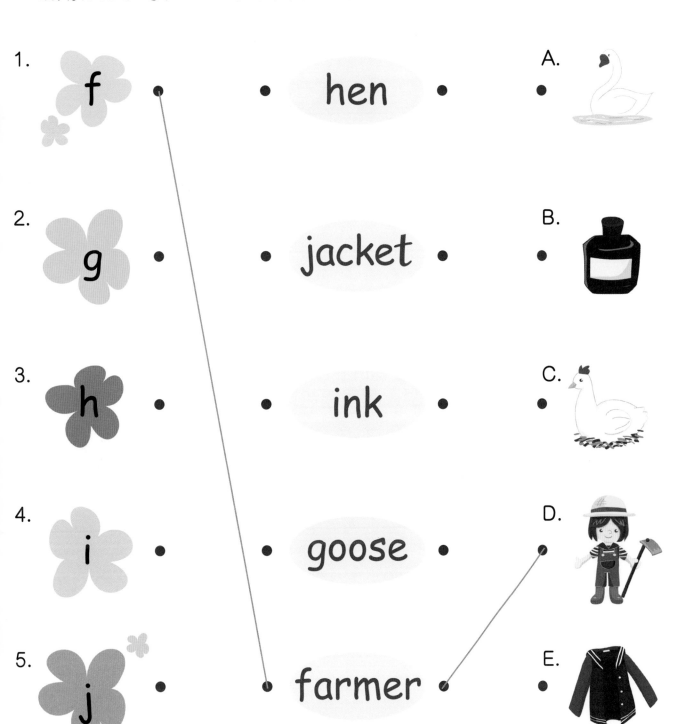

1. f •   • hen •   A.

2. g •   • jacket •   B.

3. h •   • ink •   C.

4. i •   • goose •   D.

5. j •   • farmer •   E.

Small Letter k
小楷字母 k

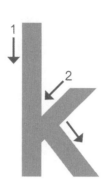

key 鎖匙

koala 樹熊

● Match the door with the key with the same letter to unlock the door.
請用線把寫有相同字母的鎖匙和門連起來。

● Let's write the letter k. 齊來寫一寫。

k	k			

Small Letter l
小楷字母 l

lamp 電燈

lemon 檸檬

● Fill in the letter " l ". 請寫小楷字母 l。

_____adder

_____ove

● Let's write the letter l. 齊來寫一寫。

l	l				

15

Small Letter m
小楷字母 m

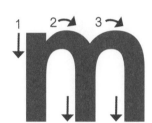

mouse 老鼠

milk 牛奶

● Help the mouse find the cheese by colouring the letter "m"s.
請把字母 m 填上顏色，帶小老鼠找到芝士吧。

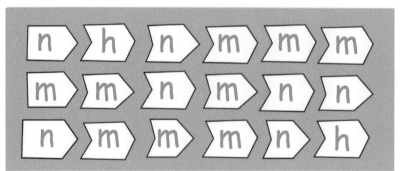

● Let's write the letter m. 齊來寫一寫。

m m				

Small Letter n
小楷字母 n

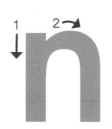

net 網

nail 釘

● Circle the things that start with the letter "n". 請圈出以字母 n 開頭的物件。

1.   2.   3.

● Let's write the letter n. 齊來寫一寫。

n	n				

答案：1. noodles   2. nests

17

Small Letter o
小楷字母 o

做得好！ 不錯啊！ 仍需加油！

octopus 八爪魚   onion 洋蔥

● Circle all the oranges with the letter "o".
請圈出所有寫上字母 o 的橙。

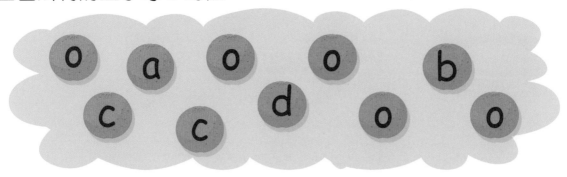

● Let's write the letter o. 齊來寫一寫。

o	o				

18

● Circle the correct picture that starts with the letter in the first column for each line.
根據左邊的字母，請圈出正確的圖畫。

1. **k**		
2. **l**		
3. **m**		
4. **n**		
5. **o**		

Small Letter p
小楷字母 p

 **p**anda 熊貓

 **p**lant 植物

Draw a line to match the picture with the correct word.
請把圖畫和正確的詞彙用線連起來。

1.

• policeman •

• pear •

2.

Let's write the letter p. 齊來寫一寫。

p	p				

20

Small Letter q
小楷字母 q

question 問題

quiet 安靜

Colour all the bubbles with the letter "q".
請把所有寫上字母 q 的泡泡填上顏色。

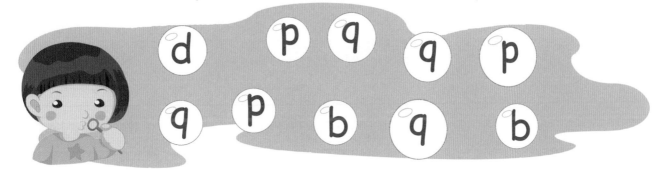

Let's write the letter q. 齊來寫一寫。

q	q			

21

Small Letter r
小楷字母 r

rainbow 彩虹

rose 玫瑰

What is this? Circle the correct word.
這是什麼？請圈出正確的詞彙。

rail

rain

rocket

ring

Let's write the letter r. 齊來寫一寫。

r	r				

答案：ring

22

Small Letter s
小楷字母 s

strawberry
士多啤梨

spoon 匙子

Fill in the letter "s". 請填寫小楷字母 s。

_____ hoes

_____ ocks

Let's write the letter s. 齊來寫一寫。

s	s				

23

Small Letter t
小楷字母 t

tree 樹

table 桌子

Circle the animals that start with the letter "t". 請圈出以字母 t 開頭的動物。

1.
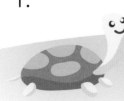

2.

3.

Let's write the letter t. 齊來寫一寫。

t	t				

24

Let's review p to t
溫習 p 至 t

Find the balloons with the letters p, q, r, s and t. Draw lines to connect them with the clown's hands.
請找出寫有字母 p, q, r, s 和 t 的氣球，然後用線把它們和小丑的手連起來。

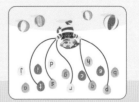

答案:

Small Letter u
小楷字母 u

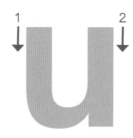

uniform 校服

unhappy 不開心

● What do you wear to school? Put a "✔" in the correct box.
你上學的時候會穿什麼？請在正確答案的 ☐ 內加 ✔。

pyjamas ☐ uniform

● Let's write the letter u. 齊來寫一寫。

u	u				

答案：uniform

26

Small Letter v
小楷字母 v

vase 花瓶

van 小型貨車

● What do you put in a vase? Put a "✔" in the correct box.
你會在花瓶裏放些什麼？請在正確答案的 ☐ 內加 ✔。

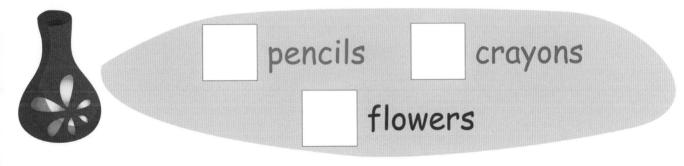

☐ pencils   ☐ crayons

☐ flowers

● Let's write the letter v. 齊來寫一寫。

v	v				

答案：flowers

27

Small Letter w
小楷字母 w

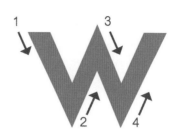

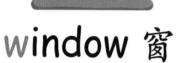

window 窗

watch 手錶

● Circle the things that start with the letter "w". 請圈出以字母 w 開頭的物件。

1.　2.　3.

● Let's write the letter w. 齊來寫一寫。

w	w				

28

Small Letter x
小楷字母 x

fox 狐狸

box 盒子

● What is the number on the box? Circle the answer.
盒子上的數字是多少？請圈出正確的答案。

one   four

six   nine

● Let's write the letter x. 齊來寫一寫。

x	x			

答案：six

29

Small Letter y
小楷字母 y

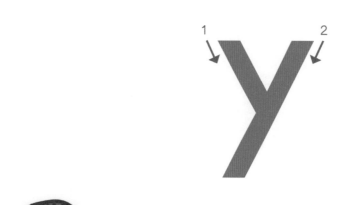

yarn 毛線　　　　yellow 黃色

- A ball of yellow yarn is used to knit into a sweater. Which sweater is it? Join all the letter "y"s and find out the answer.
請把字母 y 用線連起來，看看這個毛線球可織成哪件毛衣。

1.

y	y	v	x	v	v
v	y	y	y	y	x
x	v	v	x	y	y

2.

- Let's write the letter y. 齊來寫一寫。

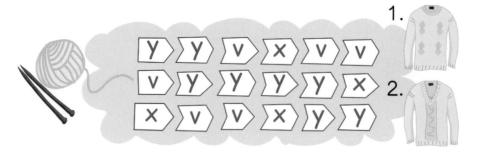

y	y				

答案：2

30

Small Letter z
小楷字母 z

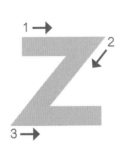

zoo 動物園

zero 零

● Trace the zigzag line to help the butterfly find the flower. Fill in the letter "z". 請沿之字線畫一畫，幫助蝴蝶飛到花朵上，然後在橫線上填寫小楷字母 z。

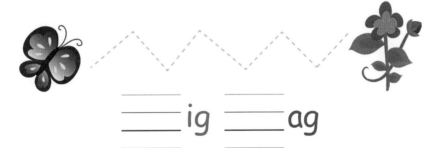

___ig ___ag

● Let's write the letter z. 齊來寫一寫。

z	z			

31

Let's review a to z
温習 a 至 z

● Fill in the missing letters in alphabetical order.
請依英文字母的順序，在星星上填寫缺少了的小楷字母。

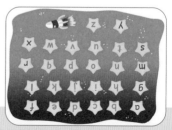

答案：